Matter

AF796856

Harcourt
SCHOOL PUBLISHERS

Orlando Austin New York San Diego Toronto London

Visit *The Learning Site!*
www.harcourtschool.com

Lesson 1

What Is the Structure of Matter?

VOCABULARY

atom
molecule
nucleus
element
periodic table

An **atom** is the smallest particle of matter that still behaves like the original matter it came from. The picture shows a hydrogen atom.

A **molecule** is made of two or more atoms. The picture shows a water molecule. It has an oxygen atom in the middle and two hydrogen atoms.

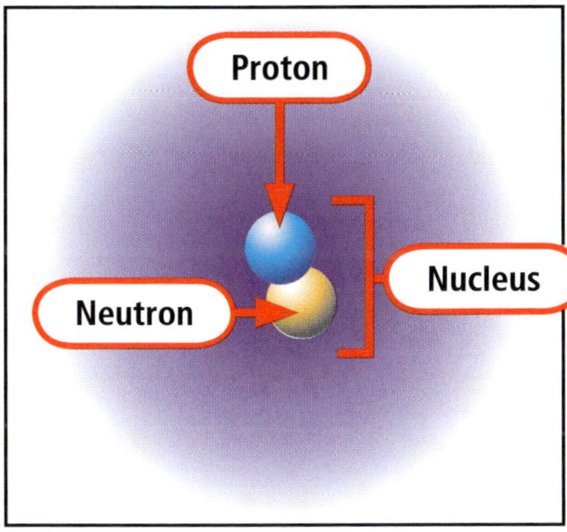

The **nucleus** of an atom is the dense area in the center. It contains protons and neutrons.

An **element** is a form of matter that contains only one kind of atom. These rings are made of the element gold.

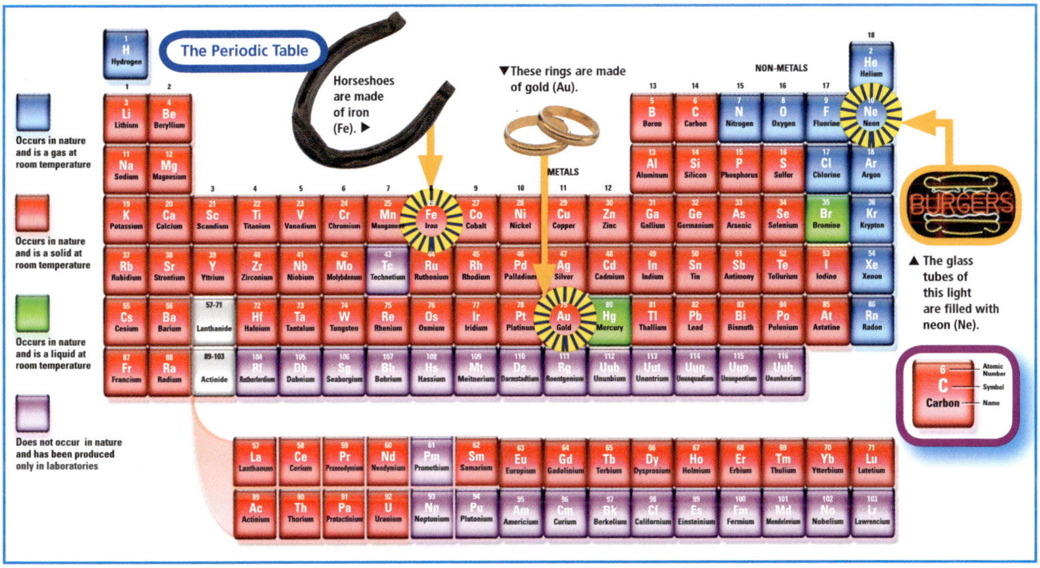

The **periodic table** is a chart of all the elements. The periodic table gives information about each element.

Focus Skill

READING FOCUS SKILL

MAIN IDEA AND DETAILS

The **main idea** is what something is about. **Details** give extra information to back up the **main idea**.

Look for **details** about the structure of atoms and molecules and how they are alike and different.

Matter

Matter is anything that takes up space and has mass. *Mass* is the amount of matter in an object. An object's mass doesn't change. Suppose you cut up an apple. The apple has the same mass before and after you cut it up.

Volume is the amount of space something takes up. Suppose you have eight ounces of water in a measuring cup. You can pour the water into a glass. It will then have a different shape. But its volume will still be 8 ounces.

◀ **Each bottle has the same volume of water.**

Atoms and Molecules

All matter is made of tiny particles called atoms. An **atom** is the smallest particle of matter that still behaves like the matter it came from.

Each atom has a dense area in the center. This is the **nucleus**. Most of an atom's mass is found in the nucleus. The nucleus has two kinds of particles inside it. Every atom has at least one *proton*. The proton has a positive electrical charge.

Most atoms also have *neutrons*. Neutrons do not have an electical charge.

Electrons travel around the nucleus. They are much smaller than protons and neutrons. Electrons have a negative charge. Usually, the number of electrons equals the number of protons.

Focus Skill **Tell how many electrons this carbon atom has.**

A carbon atom has six ▶ protons and six neutrons.

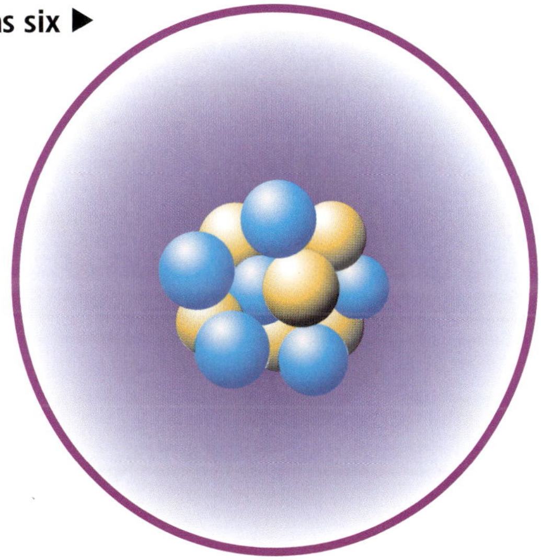

Two or more atoms joined together make up a **molecule**. Water is made up of molecules. Each water molecule is made of one oxygen atom and two hydrogen atoms. Oxygen and hydrogen are both gases at room temperature. But when they come together to form water, they make something new. Water is a liquid at room temperature.

Focus Skill **Explain the difference between an atom and a molecule.**

Waterfalls are made up of millions of water molecules. ▼

◀ **Diamonds, graphite, and coal are all forms of the element carbon.**

Elements

An **element** is matter composed of just one kind of atom. For example, hydrogen gas is made of only hydrogen atoms, so it is an element. Water is made of two kinds of atoms, hydrogen and oxygen. So water is not an element.

Scientists have identified 116 elements. Some are familiar to you. For example, pencil "lead" is actually graphite, a form of carbon. Carbon is element number 6. The neon in neon signs is also an element. Gold, silver, and iron are familiar elements that are metals.

Focus Skill **Tell how an element is different from other kinds of matter.**

Scientists organize the elements with the **periodic table**. This is a large chart that arranges elements by their *atomic numbers*. An element's atomic number is the number of protons in its nucleus.

Each box in the periodic table contains information about the element. The chart below shows each element's name and symbol. Symbols have one or two letters. The symbol for hydrogen (1) is H. The symbol for carbon (6) is C. The symbol for neon (10) is Ne. The boxes also show each atom's atomic number.

Focus Skill **Tell how elements are placed on the periodic table.**

Periodic Table

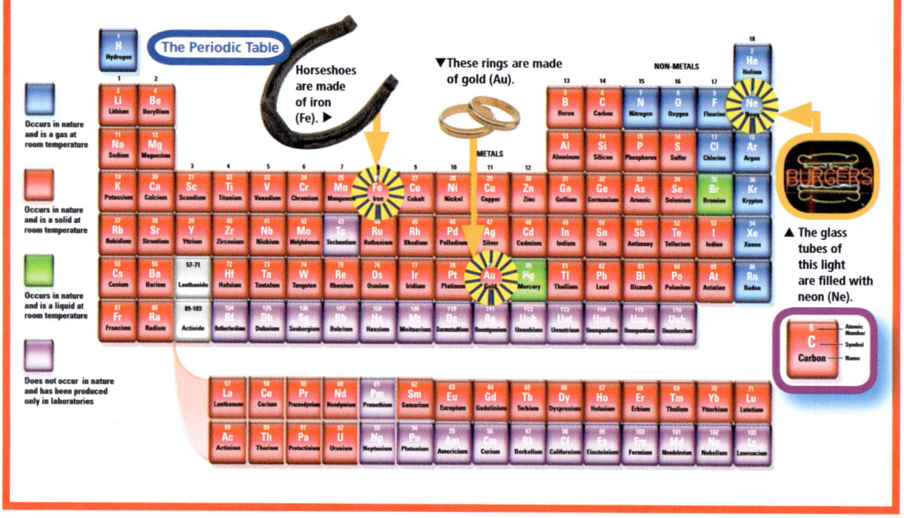

Solids, Liquids, and Gases

Some of the boxes on the periodic table have different colors. The colors show the different *states of matter*. We usually talk about three states: solids, liquids, and gases.

A solid has its own shape and volume. The particles in a solid are close together and do not move much.

A liquid has its own volume, but takes the shape of its container. Its particles are farther apart and move a little.

A gas takes the shape and volume of its container. Its particles are far apart and move fast.

▲ **Particles in solids, liquids, and gases are arranged differently.**

Review

Focus Skill

Complete these main idea sentences.

1. Two or more ______ join together to form a molecule.
2. Positively charged ______ are in an atom's nucleus.

Complete these detail sentences.

3. Matter made of only one kind of matter is a(n) ______.
4. Elements are organized in the ______ ______.

Lesson 2

What Are Physical Properties and Changes?

VOCABULARY

physical change
density
mixture
solution

A **physical change** is a change in which the substance stays the same, but its form changes. A change in state is a physical change.

Density is a measure of how concentrated matter is in an object. Less-dense liquids float on denser ones.

A **mixture** is a combination of different substances. The substances don't change. They can be separated from the mixture and be the same as they were before.

A **solution** is a mixture that has all parts mixed together evenly. It looks the same throughout.

READING FOCUS SKILL

COMPARE AND CONTRAST

To **compare and contrast** is to show how things are alike and different.

Look for ways that mixtures and solutions are alike and different.

Changing States of Matter

Water can change from one state to another. It can be a liquid, solid, or gas. Matter changing from one state to another is called a **physical change**. Many physical changes are caused by heating or cooling.

Water Changes

When its temperature changes, water can change from a solid to a liquid to a gas and back again.

Physical changes cause a change in the form of matter, but not the matter itself. Look at the pictures below.

The pitcher in the middle has liquid water. If you *freeze* this water, it becomes ice. Ice is the solid form of water. If you *melt* the ice, it goes back to liquid water. Water freezes at 32°F. It also melts at this temperature.

Look at the other side of the pitcher. You can *boil* the water at 212°F. At that temperature, water changes into water vapor. *Water vapor* is the gas form of water. Water vapor *condenses*, or changes back to a liquid, at the same temperature.

Through all these changes, the water is still water. It just changes states.

Focus Skill **Explain what a physical change is and give an example of one.**

Melting and Boiling Points

Ice melts at 32°F. This is the same as 0°C. Ice always melts at this temperature. It is called the melting point of water.

Water always boils at 212°F, or 100°C. This is called the boiling point of water.

Melting point and boiling point are *physical properties* that describe matter. Another physical property is something's state of matter at room temperature.

The graph below shows melting and boiling temperatures of different kinds of matter. Suppose you have an unknown substance. You can find its melting and boiling points. They can tell you what kind of matter it is.

Notice that the melting point is always lower than the boiling point. ▼

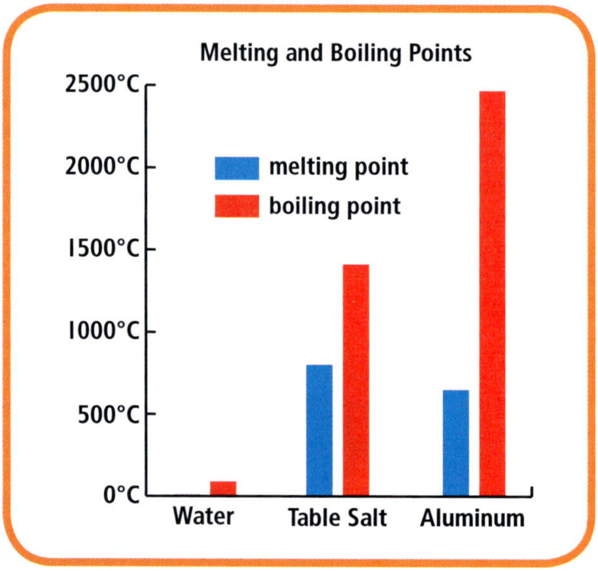

Density

To find density, you divide the mass by the volume. The three cubes in the picture all have the same volume. They are all the same size. But they have different amounts of matter. So they have different densities.

Density is a physical property of matter. **Density** tells how concentrated its matter is. Every kind of matter has its own density.

Focus Skill **The copper cube below has the greatest mass. The wood cube has the least mass. Tell what cube has the greatest density.**

Mixtures and Solutions

A **mixture** is a combination of two or more different substances. The substances in a mixture keep their own properties. They do not permanently combine. You can separate the substances and get back what you started with.

Suppose you put salt in water. This is a mixture. The water keeps its properties. And you can taste the salt. You can then boil the mixture. The water will pass into the air as water vapor. The salt will be in the bottom of the container. You get back what you started with.

You can also make a mixture with sugar and water. The sugar will seem to disappear. This is a solution. A **solution** is a mixture in which all the parts are mixed evenly.

Focus Skill **Explain why the salad dressing and the sugar and water are mixtures.**

▲ You get a mixture by combining substances.

Other Physical Changes

You see physical changes every day. Cutting paper is a physical change. Crushing a can is a physical change. Bending a piece of cardboard is a physical change.

You can also see physical changes in nature. A puddle evaporates. Your breath condenses on a cold day. A tree falls down.

Focus Skill **Explain why crushing a can is a physical change.**

◀ **Cutting paper is a physical change.**

Review

Focus Skill **Complete these compare and contrast statements.**

1. A mixture in which the parts are mixed evenly is called a ______.
2. Substances in a ______ aren't permanently combined.
3. An object's ______ is a physical property that relates mass and volume.
4. A substance's melting point is the same as its ______ point.

Lesson 3

What Are Chemical Properties and Changes?

VOCABULARY

combustibility
reactivity

Combustibility is a measure of how easily something will burn. When something burns, it is combining with oxygen.

A **compound** is made of two or more elements. Table salt is made of the element sodium and the element chlorine.

The ability of a substance to go through a chemical change is its **reactivity**. Chemical changes form new substances.

READING FOCUS SKILL
CAUSE AND EFFECT

A **cause** is what makes something happen. An **effect** is what happens.

Look for **causes** of chemical changes and how they **effect** matter.

Chemical Changes

Chemical changes are different from physical changes. A chemical change forms new substances. Look at the burnt marshmallows in the picture below. They have a black substance on them. They didn't have that before. That black substance is a new substance. The black substance was formed during a *chemical reaction*. A **compound** is a substance made of two or more different elements.

▲ The marshmallows go through a chemical change.

Burning has a special name in science. It is called *combustion*. The **combustibility** of a substance is a measure of how easily it burns. When something burns, it combines with the element oxygen.

Corrosion is another chemical change. Corrosion happens to metals. For example, iron combines with oxygen in the air to form rust.

Corrosion forced the Statue of Liberty to close for years for repairs. The statue is made of iron and copper. These metals combined with oxygen and formed rust. The rust was a new substance. The rust made the statue weak.

Explain what the measure of combustibility tells you.

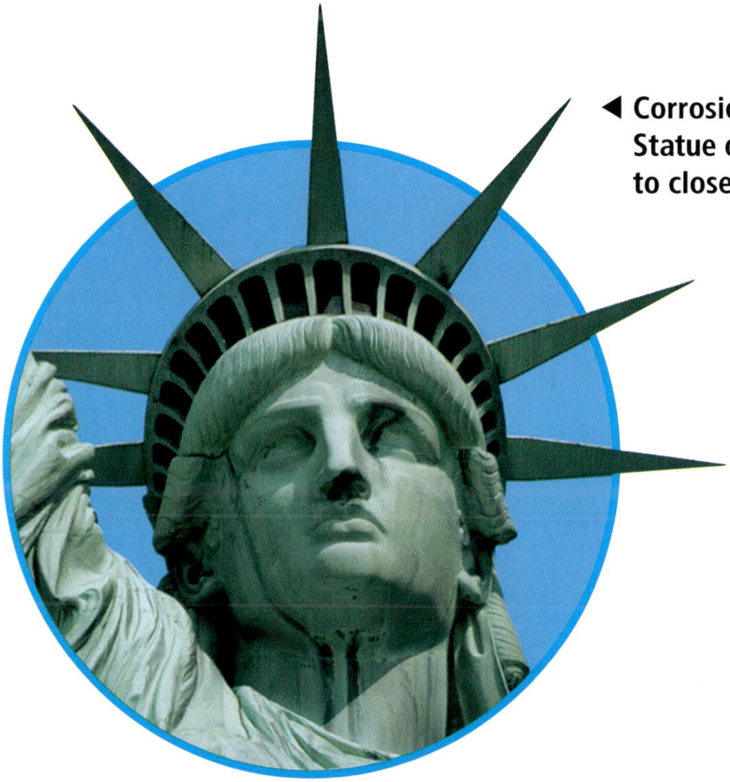

◀ **Corrosion forced the Statue of Liberty to close.**

Sometimes hydrogen peroxide is put on a cut. Gas bubbles form. The peroxide reacts with blood and releases oxygen gas. These bubbles come from a chemical reaction. The blood reacts with peroxide. The ability of a substance to react, or go through a chemical change, is called **reactivity**.

A change in color can also tell you that a chemical change has occurred. When you watch fireworks, you are watching chemical changes. If you see blue, you are watching copper burn. Burning aluminum produces a white color. Strontium burns red and barium burns green.

Focus Skill **Name three effects of chemical changes.**

Fireworks result from chemical changes. ▶

Conservation of Matter

Chemical changes do not make new matter. They change matter. The mass of the substances you had before the chemical change will always equal the mass of the substances you have after the chemical change. This is called the *law of conservation of matter.*

It might be hard to find all the masses. For example, a burned marshmallow has less mass after it is burned. Some of its mass went into the air.

The lightstick has the same mass before and after the chemical change.

Review

Focus Skill

Complete these cause and effect statements.

1. A substance burns when it combines with ______.
2. When iron combines with oxygen in the air, ______ forms.
3. Peroxide reacts with blood and releases ______ ______.
4. A change in color can be caused by a ______ ______.

GLOSSARY

atom (AT•uhm) The smallest particle that still behaves like the original matter it came from.

combustibility (kuhm•buhs•tuh•BIL•uh•tee) A measure of how easily a substance will burn.

compound (KAHM•pound) A substance made of two or more different elements.

density (DEN•suh•tee) The measure of how closely packed an object's atoms are.

element (EL•uh•muhnt) Matter made up of only one kind of atom.

mixture (MIKS•chuhr) A combination of two or more different substances.

molecule (MAHL•ih•kyool) Two or more atoms joined together.

nucleus (NOO•klee•uhs) A dense area in the center of an atom that contains protons and neutrons.

periodic table (pir•ee•AHD•ik TAY•buhl) A chart that scientists use to organize the elements.

physical change (FIZ•ih•kuhl CHAYNJ) A change in which the form of a substance changes but the substance still has the same chemical makeup.

reactivity (ree•ak•TIV•uh•tee) The ability of a substance to go through a chemical change.

solution (suh•LOO•shuhn) A mixture in which all the parts are mixed evenly.